HOW HUMANITY COULD COLONIZE TITAN
The Next Earth?

An Analytical Odyssey into Terraforming, Exploration, Science, Mystery, and Promise of Saturn's Most Intriguing Moon

Scott W. Diego

Copyright © Scott W. Diego, 2024.

All rights reserved. No part of this publication may be reproduced, distributed, or transmitted in any form or by any means, including photocopying, recording, or other electronic or mechanical methods, without the prior written permission of the publisher, except in the case of brief quotations embodied in critical reviews and certain other noncommercial uses permitted by copyright law.

Table of Contents

Introduction..3
Chapter 1: Discovering Titan – A World Beyond Earth.. 6
Chapter 2: Titan's Landscape and Atmosphere...... 14
Chapter 3: The Science Behind Titan's Methane Cycle.. 23
Chapter 4: Surviving Titan's Hostile Environment.. 31
Chapter 5: Terraforming Titan – Possibilities and Challenges... 38
Chapter 6: The Potential for Life on Titan................ 47
Chapter 7: NASA's Dragonfly Mission – A New Era of Exploration.. 57
Chapter 8: Titan as Humanity's Next Home – Infrastructure and Resource Utilization.................... 67
Chapter 9: The Ethical and Practical Implications of Colonizing Titan... 76
Chapter 10: The Future of Human Exploration – Titan and Beyond... 87
Conclusion... 98

Introduction

Imagine a world tucked away in the reaches of our solar system, where rivers and lakes carve paths through a dense, hazy atmosphere, and mountain ranges stand under a dim sky that never brightens past twilight. This world, both familiar and alien, is Titan—Saturn's largest moon. With its vast reserves of liquid methane and landscapes sculpted by geological forces unknown to Earth, Titan presents an unlikely blend of features that mirror aspects of our planet, yet diverge in every essential way. It's a place that defies the imagination, where mountains are formed not from rock but from ice, and where the atmosphere teems with nitrogen, much like our own, but with a total absence of oxygen.

This otherworldly allure of Titan has captured the curiosity of scientists and dreamers alike. In an era where humanity contemplates its first steps beyond Earth, Titan offers a tantalizing promise: a moon shrouded in mystery, with the potential to become more than just a far-off curiosity. Unlike Mars or our own moon, Titan stands out due to its thick atmosphere, its robust chemical cycles, and the protections it affords against cosmic radiation, thanks to the embrace of Saturn's magnetic field. Each of these qualities has prompted experts to envision what it might take to transform this icy moon into a sanctuary for human life, to craft a place where humanity might one day thrive despite the formidable obstacles.

Yet the path toward making Titan habitable isn't a straightforward one. The challenges are formidable: a landscape dominated by extreme cold, a complete absence of breathable oxygen, and the omnipresence of flammable gases. This raises fundamental questions—how could we reshape Titan's environment to make it livable, and what costs might accompany such an endeavor? Could we harness the elements already present, or would the task require importing resources on a scale that defies current human capability?

This book takes readers on a journey through these questions and beyond. It explores the science of Titan's unique chemistry, the practical challenges and opportunities of exploration, and the potential for terraforming a world unlike any other. Alongside the scientific inquiry lies a thread of mystery, a sense of the unknown that adds a new dimension to our drive for exploration. For with each piece of knowledge we uncover, new questions arise about the secrets Titan may be holding.

In the pages that follow, we'll delve into these themes, unpacking the scientific complexities, examining humanity's aspirations, and confronting the moral and practical implications of reshaping a distant world. Titan's promise, its obstacles, and its potential to change the course of human destiny make it more than just a scientific target; it's a symbol of what might be possible when human curiosity and ingenuity venture to the farthest reaches of our cosmic neighborhood.

Chapter 1: Discovering Titan – A World Beyond Earth

Titan, Saturn's largest moon, occupies a distinct place in our solar system as both a familiar and an alien world. Larger than Mercury and second in size only to Jupiter's Ganymede, Titan possesses characteristics that make it more than a mere satellite. Wrapped in a dense, nitrogen-rich atmosphere, it stands as one of the most Earth-like locations beyond our planet, though its surface hides strikingly different materials and conditions. The allure of Titan isn't just in its size but in its intriguing composition: an icy shell, a vast subterranean ocean, and an atmosphere thick enough to support weather patterns, rain, and even lakes. These unique elements make Titan a captivating target for exploration, drawing scientists and explorers into a long-standing pursuit to understand and perhaps one day visit or even colonize this distant moon.

Our first glimpses into Titan's mysteries began with the Voyager missions in the early 1980s. As the Voyager spacecraft soared past Saturn, it captured images of Titan that hinted at its clouded surface and suggested an atmosphere unlike that of any other moon in the solar system. While Voyager couldn't penetrate Titan's thick haze to reveal surface details, it confirmed what many had speculated: Titan was unique, enveloped in an atmosphere dominated by nitrogen, similar to Earth's but much denser, with a suffocating lack of oxygen. The possibilities hinted at by Voyager fueled an intensified interest in exploring this moon more thoroughly.

This curiosity found new life with the Cassini-Huygens mission, a joint venture between NASA, the European Space Agency (ESA), and the Italian Space Agency. Cassini, launched in 1997, reached Saturn and began orbiting it in 2004, providing us with an unparalleled look at Titan. Its companion, the Huygens probe, was designed specifically to touch down on Titan's surface, marking humanity's first landing on a moon of Saturn. As Huygens descended through Titan's atmosphere in January 2005, it transmitted data for 90 minutes, capturing sounds and images that revealed an alien landscape with features reminiscent of Earth's—river channels, possible shorelines, and a surprisingly solid surface below its hazy skies.

From its position in orbit, Cassini offered an extended view, conducting numerous flybys that revealed an astounding level of detail. Radar mapping revealed lakes and seas concentrated near Titan's poles, filled not with water but with liquid methane and ethane, hydrocarbons that behave like Earth's water under Titan's frigid conditions. Titan's rivers and lakes confirmed that a complex cycle, similar in ways to Earth's hydrological cycle, was at play, though driven by methane instead of water. Cassini also uncovered evidence of vast dunes near the equator, composed of hydrocarbon particles that act like grains of sand, shaped by Titan's winds into familiar yet alien patterns.

Cassini's extended mission lasted until 2017, and during those years it fundamentally transformed our understanding of Titan. Scientists learned that Titan's crust is made up of water ice, rigid and rock-like in its extreme cold, and that beneath this shell lies a global ocean, likely of liquid water mixed with ammonia. This ocean, hidden under a hundred kilometers of ice, has only deepened the mystery of whether Titan could harbor life or perhaps support human habitats in the future. Cassini also detected cryovolcanic activity—volcanoes that release icy slush instead of molten lava—suggesting that Titan's interior might be more active than previously thought.

Through these missions, Titan went from a hazy enigma to one of the most scientifically valuable targets in the solar system. Each discovery has added to a growing sense that Titan, while challenging and hostile, holds secrets that might one day unlock new possibilities for life beyond Earth. With its lakes of methane, its icy mountains, and its hidden ocean, Titan remains a world full of surprises, inviting us to look closer, question further, and dream beyond the familiar boundaries of our own planet.

Titan's dense atmosphere sets it apart as a prime candidate for exploration in ways that few other celestial bodies can rival. In a solar system filled with barren, airless moons and thinly veiled planets, Titan's atmosphere stands as a protective and dynamic barrier, rich with nitrogen and layered with complex hydrocarbons. This thick atmosphere not only insulates the surface but also provides a level of protection from cosmic and solar radiation, sparing Titan from the intense bombardment that other moons and planets face. Unlike Mars, which has a thin atmosphere unable to protect surface-dwelling organisms or potential explorers, Titan's atmospheric density adds a significant layer of shielding, making it comparatively safer for future human missions.

With a composition primarily of nitrogen, similar to Earth's, Titan's atmosphere is a unique feature in the solar system. This nitrogen-rich environment means the atmospheric pressure on Titan's surface is around 1.5 times that of Earth, creating conditions under which humans could theoretically walk around without a pressurized suit—an unimaginable advantage compared to Mars or the Moon. This makes Titan not only a fascinating scientific target but also one of the more physically accessible environments for exploration, as future visitors would need thermal protection against the cold but wouldn't have to contend with the vacuum of space or the brutal pressure of Venus.

Titan's atmosphere also brings with it weather patterns shaped by methane clouds, which create rain, rivers, and lakes through a cycle similar to Earth's water cycle. In Titan's world, methane takes on the role water does on Earth, forming clouds, condensing, and raining down to fill its lakes and seas. This cycle of methane rainfall and evaporation indicates an active and evolving surface, with rivers that carve through icy terrain and vast seas that ebb and flow with changing temperatures. It's a landscape shaped by processes familiar yet strikingly different, making Titan a place where Earth-based science can be applied to alien phenomena in ways that expand our understanding of planetary systems.

Another Earth-like feature is Titan's seasonal shifts, influenced by Saturn's 29-year orbit around the Sun. As Titan journeys through seasons similar to Earth's, albeit much slower, its climate changes in ways that allow scientists to observe and model atmospheric dynamics that may even resemble early Earth's own atmospheric evolution. Titan's dense atmosphere, rich in organic molecules, also hints at prebiotic chemistry—chemical processes that could be similar to the early stages of life on Earth. While life as we know it is unlikely on Titan's surface, its atmosphere and potential for liquid water below the icy crust raise questions about what forms of life, if any, could exist in such an environment. This possibility of unique, methane-based or subsurface life is tantalizing, adding to Titan's scientific appeal.

The conditions on Titan present an intriguing middle ground between Earth and other planets in terms of habitability, making it an ideal destination for studying how life might arise in different chemical and environmental conditions. Titan's atmosphere and Earth-like features thus not only make it a natural scientific target but also a potential stepping stone in humanity's quest to understand life's potential in the universe, challenging us to rethink what "habitable" means in entirely new terms.

Chapter 2: Titan's Landscape and Atmosphere

Beneath Titan's thick, orange-tinted haze lies a landscape that appears almost Earth-like in form, yet is fundamentally alien in substance. This moon of Saturn, though frigid and far from the Sun, has carved-out landscapes shaped by elements that behave nothing like the water, soil, and rocks of Earth. Titan's surface reveals a mosaic of methane lakes, seas, and towering mountains—all of which contribute to its reputation as one of the solar system's most geologically complex moons.

Titan's lakes and seas, concentrated near the polar regions, are filled with liquid methane and ethane, hydrocarbons that can exist in a liquid state at Titan's surface temperatures of around -180 degrees Celsius (-290 degrees Fahrenheit). The largest of these bodies, Kraken Mare, stretches as wide as the state of Texas and plunges hundreds of meters deep. Neighboring seas, like Ligeia Mare and Punga Mare, create a network of liquid bodies fed by rivers that wind across Titan's icy terrain, mimicking the familiar process of river-fed lakes on Earth. This methane-based hydrology is a defining feature of Titan's surface, cycling through evaporation, condensation, and precipitation, just as water does on Earth, but within an entirely different chemical framework.

These methane rivers and lakes are fed by seasonal rains, creating a dynamic surface that shifts and reshapes over time. Methane clouds form in Titan's atmosphere, producing periodic rainstorms that fall to the surface and replenish the lakes. This cycle is thought to shape valleys and channels across Titan's icy crust, forming landscapes that seem familiar but are profoundly foreign. While Earth's rivers are cut by liquid water flowing over rock and soil, Titan's channels are carved by liquid hydrocarbons moving across frozen terrain composed of water ice and hydrocarbon "sand."

Beyond its lakes and seas, Titan also boasts vast dunes and mountain ranges. Near the equatorial regions, the surface is dominated by long stretches of sand dunes, formed not of silica like Earth's sands but of organic compounds that settle on Titan's surface, resembling plastic grains. Winds driven by Titan's thick atmosphere sweep these particles into dunes hundreds of meters high, creating patterns across the surface that give the region a desert-like appearance. Meanwhile, mountain ranges rise up in areas of tectonic or cryovolcanic activity, where icy material pushed from below the surface creates jagged formations. These mountains, primarily composed of water ice hardened by Titan's extreme cold, rise like rocky peaks on Earth, though their composition is fundamentally different.

At the root of this alien terrain lies Titan's atmosphere, a dense shroud that blocks much of the sunlight and bathes the moon in a perpetual twilight. This atmosphere, rich in nitrogen but virtually devoid of oxygen, creates an environment that appears familiar in some respects but ultimately defies Earthly conditions. Titan's nitrogen-based atmosphere resembles Earth's, but instead of the oxygen that makes up 21% of Earth's atmosphere, Titan's is laced with methane and hydrogen, giving rise to a cocktail of organic compounds. These compounds react under the influence of weak sunlight and cosmic rays, forming the thick orange haze that obscures Titan's surface from view.

This haze, created by a complex mix of hydrocarbons, including acetylene and ethane, blocks most visible light and allows only faint twilight to reach the ground. It also creates Titan's characteristic orange hue, a surreal contrast to the pale yellow and blue of Saturn in the distance. At high altitudes, ultraviolet light from the Sun breaks down methane, leading to a cascade of chemical reactions that produce a variety of organic molecules. Over time, these compounds descend to the surface, accumulating as hydrocarbon "snow" or settling into dunes, contributing to the moon's distinct appearance.

Titan's dense atmosphere and active surface make it a place of striking contrasts, where landscapes that look like familiar Earth features—oceans, rivers, mountains—are entirely reshaped by a different chemistry. This world, with its methane lakes and icy mountains, offers a glimpse of what a planetary body might look like if governed by the same geological processes as Earth but driven by substances that are toxic and volatile by our standards. In the thick haze, beneath skies of twilight, Titan's surface is a frontier shaped by elements that challenge our understanding of how landscapes are formed, offering scientists and explorers alike a rare chance to study a world that reflects yet redefines our concept of a habitable landscape.

Titan's environment is shaped by extreme cold and a sunlight-starved atmosphere, conditions that create challenges for any notion of human exploration, let alone settlement. The sun is nearly 1.5 billion kilometers away, so distant that light reaching Titan is only about one percent as intense as on Earth. Even at midday, Titan's skies resemble the deep twilight after Earth's sunset, casting a dim, shadowed light across its surface. Temperatures plummet to around -180 degrees Celsius (-290 degrees Fahrenheit), a frigid climate where methane and ethane remain liquid, while water exists only in the form of solid, rock-like ice. This relentless cold pervades Titan, limiting the potential for any familiar Earth-based life and demanding extreme adaptations for human technology and habitats.

This twilight world, while harsh, is not entirely static. Titan has an active geology, one that hints at forces beneath its frozen crust capable of shaping the landscape over millennia. One of the most fascinating of these geological processes is cryovolcanism, a phenomenon where "ice volcanoes" spew not lava, as on Earth, but a mixture of water, ammonia, and methane in slushy eruptions. These cryovolcanoes are thought to form as Titan's interior heat and pressure push liquid water and other chemicals from the subsurface ocean to the surface. When this frigid liquid reaches Titan's surface, it quickly freezes into solid "lava," creating unique mountainous formations that stand as evidence of past eruptions. This process adds to the moon's rocky landscape, with cryovolcanic features appearing like jagged peaks, built from the frozen remnants of once-flowing icy slush.

Below Titan's rigid crust lies a subsurface ocean, a hidden world that scientists believe could hold more liquid water than all of Earth's oceans combined. This ocean, shielded by layers of water ice up to 100 kilometers thick, is likely mixed with ammonia, which acts as an antifreeze, keeping the water from freezing solid despite the bitterly cold temperatures. This subsurface ocean, separated from the surface by a crust of crystalline ice, is one of the most tantalizing aspects of Titan for astrobiologists, as it raises the possibility of life existing in conditions similar to those found around Earth's hydrothermal vents. These deep-sea environments on Earth support ecosystems driven by chemical reactions rather than sunlight, a model that could theoretically apply to Titan's hidden ocean as well.

On the surface, Titan's landscape is made up of an unusual blend of water ice, frozen so solid that it forms rock-like formations, and hydrocarbon sands derived from the organic compounds raining down from the atmosphere. These grains of hydrocarbon settle over time, accumulating in vast dune fields near the equator, creating a landscape that resembles a desert but is made of material that, in Earth's environment, would be considered plastic-like rather than mineral-based. This "sand" interacts with Titan's methane rivers and lakes, adding to the distinct appearance of the landscape. With each new discovery, scientists are piecing together a picture of a world where icy mountains and hydrocarbon sands create features that are both eerily reminiscent of Earth and fundamentally different.

The combination of extreme cold, low sunlight, and alien materials makes Titan a challenging environment for exploration, one that would test the limits of human technology and endurance. Any future missions would require specialized equipment designed to operate at temperatures well below what is typical for Earth-based technology. Yet, in these challenges lie opportunities for discovery. Titan's surface, shaped by cryovolcanism, methane rain, and an active atmosphere, offers a unique laboratory for studying how planets and moons evolve under conditions vastly different from our own. As we contemplate humanity's future in space, Titan remains a frozen frontier, one that invites us to adapt and innovate, to push the boundaries of exploration into the twilight of Saturn's most intriguing moon.

Chapter 3: The Science Behind Titan's Methane Cycle

Titan's landscape is dominated by vast bodies of liquid methane and ethane, a phenomenon unique in our solar system. These hydrocarbon lakes, rivers, and seas mark Titan as a place where liquid flows on the surface, carving out a landscape with features that appear almost Earth-like. At the center of this network lies Kraken Mare, Titan's largest sea, sprawling over an area comparable to the size of Texas. Kraken Mare is not only immense in surface area but also reaches depths of over 300 meters, plunging into Titan's icy crust in a body of liquid that is entirely alien to Earth's water-based hydrology. Alongside Kraken Mare, smaller seas like Ligeia Mare and Punga Mare form a collection of liquid bodies mostly concentrated near Titan's polar regions, suggesting that the poles play a crucial role in Titan's unique liquid cycle.

Titan's lakes and seas aren't filled with water but rather with methane and ethane, hydrocarbons that can exist as liquids at Titan's surface temperature of around -180 degrees Celsius. This creates a hydrological cycle based on methane, where liquid methane fills Titan's lakes and seas, evaporates into methane clouds, and rains down again in a cycle much like Earth's water cycle. This process, though operating on a different substance, leads to familiar features like rivers and deltas. Methane rains, often sporadic but intense, form channels and valleys, etching their way across Titan's frozen surface and feeding the lakes and seas below. These rivers, some stretching for hundreds of kilometers, bear an uncanny resemblance to Earth's river systems, meandering through the icy terrain and depositing hydrocarbons into Titan's liquid basins.

Methane's role in Titan's environment extends beyond the lakes and rivers to the thick clouds that hover in its atmosphere. In Titan's lower atmosphere, methane vapor condenses to form clouds, sometimes accumulating in massive formations that release torrents of methane rain.

This methane rainfall is one of the few processes beyond Earth that actively shapes a planetary surface with liquid erosion. When methane rains down, it interacts with Titan's icy and hydrocarbon-rich crust, further eroding the landscape and contributing to the formation of rivers and valleys. In areas with more intense rainfall, these features evolve into complex networks of river valleys, deltas, and lakes that fill and drain with Titan's shifting seasons.

Seasons on Titan follow Saturn's 29-year orbit, so each season lasts over seven Earth years. As Titan's poles tilt closer to or farther from the Sun, methane lakes and seas experience seasonal changes, with some lakes appearing to shrink or shift as the methane levels fluctuate. These changes, observed by the Cassini mission, suggest that methane lakes might partially evaporate and condense in tune with Titan's long seasons, redistributing methane in a cycle that parallels Earth's seasonal patterns.

The methane cycle on Titan represents a complete, closed-loop system, with methane behaving in the same ways that water does on Earth: collecting in bodies, evaporating, and returning as precipitation. This cycle not only shapes Titan's surface but also offers insight into how a planetary body can maintain liquid on its surface without the involvement of water. The methane lakes, rivers, and seas are central to understanding Titan's complex environment, offering scientists a natural laboratory to study atmospheric and surface interactions under conditions that mirror early Earth but follow an entirely different chemistry. As we seek to understand the potential for life in extreme environments, Titan's methane cycle reveals a world shaped by familiar processes, yet driven by elements and conditions far removed from anything we know on Earth.

Titan's frigid climate, with average surface temperatures around -180 degrees Celsius (-290 degrees Fahrenheit), creates an environment where methane, which would be a gas under Earthly conditions, exists as a stable liquid. This extreme cold defines nearly every aspect of Titan's hydrology and atmospheric chemistry, allowing for the unique presence of liquid methane lakes, rivers, and seas. Under Titan's low-temperature conditions, methane behaves much like water does on Earth, moving through a cycle of evaporation, condensation, and precipitation. This cycle creates bodies of liquid methane that carve Titan's surface, but it also presents a unique opportunity for future explorers to consider methane as a potential resource.

One of the most intriguing prospects is the potential to harness Titan's abundant methane for energy. On Earth, methane is commonly used as a fuel, its high energy density making it valuable for heating and electricity generation. In a Titan habitat, methane could be similarly harnessed, potentially providing a stable and abundant energy source. While burning methane in Titan's oxygen-free atmosphere would not be possible, using it within controlled environments where oxygen is supplied could create efficient energy systems for heating, electricity, or even propulsion.

Moreover, methane could serve as a fundamental building block in the development of essential materials for human infrastructure. Methane can be chemically transformed into a range of hydrocarbons, which can then be used to produce plastics and other construction materials. In this way, the hydrocarbon-rich environment of Titan presents not just a fuel source but also the raw materials for creating everything from basic shelter components to complex structures. In an environment as remote as Titan, such in-situ resource utilization could be essential for reducing reliance on supply chains from Earth, enabling the development of more self-sufficient human habitats.

The cold temperatures on Titan also mean that methane can be stored easily, remaining in a stable liquid state without complex refrigeration, which would otherwise be necessary under warmer conditions. This stability allows for efficient storage and transport of methane within future Titan-based habitats or vehicles. Additionally, the abundance of methane and other hydrocarbons on Titan could allow for the production of fuel for spacecraft, potentially turning Titan into a valuable refueling station for missions deeper into the solar system. A spacecraft designed to burn methane and oxygen could refuel using Titan's methane lakes in combination with oxygen extracted from water ice, opening the possibility of Titan becoming a hub for long-duration missions.

The potential uses of Titan's methane-rich environment underscore its unique value as a location not just for exploration but for resource-based sustainability. By harnessing local resources like methane, future missions to Titan could minimize the need to transport fuel and construction materials from Earth, allowing human presence to become more viable over the long term. In a place where methane flows like water, humanity might find one of its most unexpected resources—a means of energy, infrastructure, and exploration that capitalizes on the alien conditions of a distant moon.

Chapter 4: Surviving Titan's Hostile Environment

Surviving Titan's intensely cold environment would require protective measures beyond anything used in human space exploration so far. At an average surface temperature of around -180 degrees Celsius, Titan's climate would freeze any unprotected equipment or human skin in seconds. For humans to operate safely in such conditions, advanced thermal suits would be essential. These suits would need to be equipped with powerful insulation, likely incorporating multi-layered thermal barriers and electrically heated elements to ensure stable warmth. Built to counteract temperatures far below what typical space suits are designed for, these thermal suits would shield explorers from the icy air while maintaining flexibility for mobility and dexterity.

Beyond individual suits, the technology supporting human survival on Titan would likely include specialized habitats with integrated temperature regulation. Unlike the habitats designed for Mars, where temperatures are cold but manageable, habitats on Titan would need to sustain internal warmth continuously, with thick walls that insulate against the external freeze. Advanced heating systems, powered by local resources like methane, could provide the steady warmth required to make Titan habitable for humans. Additionally, Titan's prolonged and deep twilight would require artificial lighting in habitats to simulate day-night cycles, essential for human circadian rhythms.

Interestingly, Titan's atmosphere offers one advantage that distinguishes it from other potential locations for human exploration: its high atmospheric pressure. At around 1.5 times the pressure of Earth's atmosphere, Titan's air pressure is similar to the feeling of being 15 meters underwater, enough to support the human body without requiring a pressurized suit. This pressure balance means that explorers wouldn't need bulky, pressurized suits like those necessary on Mars or the Moon. Instead, lightweight thermal suits, fitted with respirators for oxygen, could allow humans to walk Titan's surface with relatively free movement. This reduced equipment burden could make exploration more flexible, especially for tasks requiring dexterity and extended mobility.

However, breathable air is absent from Titan's nitrogen-rich, oxygen-devoid atmosphere, so respirators would be essential. These respirators would supply pure oxygen, filtering out methane, nitrogen, and other gases that are either toxic or unusable by humans. Since Titan's atmosphere also lacks significant amounts of carbon dioxide, which humans typically exhale, these respirators would need systems to scrub and recycle exhaled oxygen, ensuring a constant and safe air supply.

While Titan's thick atmosphere provides a shield against cosmic and solar radiation, eliminating the need for radiation-shielded suits or habitats, the thermal requirements remain an ongoing challenge. Maintaining warm conditions within both personal suits and larger habitats would be a continuous, resource-intensive task. In a place where methane is abundant but temperatures are well below freezing, human technology would need to adapt, merging thermal protection with breathable air systems to sustain life in one of the most extreme environments imaginable.

Through this balance of protective suits, oxygen supply, and self-heating habitats, humans could find a way to exist within Titan's atmospheric pressure and unique conditions, opening the door to exploring a world that defies many conventional standards of habitability.

On Titan, survival hinges not only on protection from the cold but also on overcoming the inhospitable composition of the atmosphere. Unlike Earth, where oxygen comprises around 21% of the air we breathe, Titan's atmosphere contains nearly zero oxygen. Instead, it is rich in nitrogen, methane, and hydrogen, creating a mixture that would be not only unbreathable but potentially lethal to humans.

To explore Titan, humans would require advanced breathing apparatuses that could supply pure oxygen, filtering out gases that are incompatible with life or dangerously flammable.

The abundance of methane and hydrogen on Titan presents another unique risk. Both gases are highly combustible in the presence of oxygen, meaning that any oxygen source introduced into Titan's atmosphere could create explosive reactions if not carefully controlled. Breathing apparatuses would need to isolate oxygen supplies entirely from the external atmosphere, ensuring that no reactive oxygen escapes into the flammable methane-rich environment. Additionally, these systems would likely incorporate robust safety measures to prevent any accidental leaks, allowing explorers to move through Titan's atmosphere without risking spontaneous combustion.

Despite the need for specialized breathing systems, Titan offers one advantage that many other celestial bodies do not: significant protection against cosmic radiation. Saturn's immense magnetosphere acts as a natural shield, enveloping Titan and deflecting much of the solar and cosmic radiation that would otherwise bombard the surface. For roughly 95% of its orbit, Titan remains within this magnetic field, offering protection comparable to Earth's natural defenses. This shielding effect is a crucial advantage over Mars or the Moon, both of which have minimal atmospheric or magnetic protection and require heavy radiation shielding for any human presence.

Moreover, Titan's thick atmosphere provides an additional layer of protection, absorbing much of the remaining radiation that penetrates Saturn's magnetosphere. The dense haze that blocks sunlight also filters out harmful cosmic rays, creating an environment where humans would not need the reinforced radiation suits necessary on less-protected planets. This double-layered shield—the combination of Saturn's magnetosphere and Titan's own dense atmosphere—gives Titan an unexpected edge in terms of human habitability, making it a rare outpost in the solar system where radiation concerns would be relatively minor.

This natural radiation barrier could simplify life on Titan, allowing explorers to focus on temperature control and breathable air rather than complex radiation shielding. With breathing apparatuses isolating oxygen and habitats safely containing both warmth and an oxygen supply, humans could move through Titan's alien landscape, equipped to face its many challenges but relieved of the radiation risk that complicates missions to most other worlds. In Titan's dark skies and methane-rich atmosphere, humanity may find a strange yet potentially manageable environment, one that offers protective qualities despite its unfamiliar dangers.

Chapter 5: Terraforming Titan – Possibilities and Challenges

The idea of terraforming—altering a planet or moon's environment to make it more suitable for human life—has long been an exciting, if speculative, area of study. In the case of Titan, terraforming would mean adjusting an atmosphere dense with nitrogen and hydrocarbons, warming a frozen landscape where temperatures plunge below -180 degrees Celsius, and creating a livable, Earth-like climate. Unlike other bodies, where thin atmospheres or radiation hazards present immediate barriers, Titan's thick atmosphere offers a starting point, giving visionaries a reason to consider whether and how its environment could be gradually transformed to support human life.

One of the most discussed concepts for warming Titan involves creating a controlled greenhouse effect. The basic principle would be to introduce gases that trap heat within Titan's atmosphere, causing a gradual rise in surface temperature. This warming could be achieved by releasing greenhouse gases like carbon dioxide or certain fluorocarbons, substances that retain heat but would not react explosively with the surrounding methane. Over time, these gases would accumulate in the atmosphere, raising Titan's temperature just enough to ease some of its most extreme conditions while preserving the unique aspects of its surface and atmosphere. The controlled nature of this greenhouse effect would be crucial, as even a slight increase in temperature would significantly alter Titan's environment.

But while this approach sounds promising, it carries considerable risks. Titan's lakes and seas are filled with methane and ethane, hydrocarbons that are liquid only because of Titan's frigid temperatures. Raising the temperature just a few degrees could disrupt this delicate balance, causing methane to evaporate more quickly and potentially destabilizing Titan's atmospheric cycle. This could trigger a chain reaction where the methane evaporates faster than it can condense, leading to the depletion of Titan's lakes and rivers. Such a shift would halt Titan's methane-based hydrological cycle, turning once stable lakes into sprawling patches of evaporation.

An even greater risk lies in the potential for Titan to transform into an ocean world if warming progresses too far. Beneath Titan's icy crust lies a hidden subsurface ocean, likely composed of liquid water mixed with ammonia. This ocean, currently kept stable by the frigid surface temperatures, could begin to melt if surface conditions warm beyond a critical threshold. As the crust thins or fractures, this subsurface ocean could rise to the surface, transforming Titan's landscape into a vast, planet-wide sea. While intriguing, this scenario would drastically alter the surface, submerging the hydrocarbon dunes, methane lakes, and icy mountains, creating an entirely new environment and potentially eliminating the unique features that make Titan so scientifically valuable.

These scenarios underscore the challenge of carefully balancing any terraforming effort on Titan. Warming the surface even slightly could dramatically alter its chemical and physical landscape, with unintended consequences that might be impossible to reverse. This delicate balance requires a deep understanding of Titan's atmospheric and geological dynamics before any attempt to introduce greenhouse gases or other warming agents. The very characteristics that make Titan intriguing—its methane lakes, cryovolcanoes, and frozen dunes—could vanish, replaced by an unpredictable, ocean-dominated world with new and unknown risks.

Terraforming Titan, if ever pursued, would demand precision, restraint, and a willingness to accept the limits of what can be achieved. Titan's atmosphere and methane cycle offer tantalizing possibilities, but they are bound by conditions far removed from Earth's. To create a habitable Titan, humanity would need not only advanced technology but an ethical awareness of the irreversible changes such an endeavor would bring. In the pursuit of a livable environment, we could risk losing the very features that make Titan one of the most extraordinary locations in our solar system.

The prospect of making Titan more habitable for human life raises a complex challenge: balancing the desire for an Earth-like environment with the need to preserve the moon's distinct, scientifically valuable features. Titan is, after all, one of the most unique bodies in the solar system, a world where methane flows like water and hydrocarbon dunes stretch across the landscape. Its methane seas, nitrogen-rich atmosphere, and icy crust make Titan a natural laboratory, offering scientists a rare opportunity to study conditions vastly different from those on Earth. Transforming Titan, while enticing, comes with the risk of permanently altering or even destroying these elements, which could limit our understanding of planetary science and the possibilities for life beyond Earth.

One approach to maintaining this balance might involve making specific regions of Titan more habitable without completely altering the moon's environment. Creating "micro-environments" or enclosed habitats could allow for controlled, Earth-like conditions in isolated areas, where humans could live and conduct research without disrupting Titan's natural state. These habitats could be engineered to maintain warmth, oxygen, and safe living conditions within their walls, while outside, Titan's atmosphere and surface would remain untouched. In this way, humans could experience Titan's landscape and conduct research on its unique properties without the need for large-scale terraforming.

Another idea would be to focus any atmospheric or temperature changes on gradual, localized modifications that would enhance human exploration without disrupting Titan's larger ecosystem. For instance, using greenhouse gases strategically in specific regions rather than spreading them across the entire moon might provide warmer microclimates that facilitate limited outdoor activities while preserving the colder methane lakes and the natural methane cycle elsewhere. This approach would mean balancing minor habitability improvements with the preservation of Titan's fragile atmospheric and geological processes. Such controlled modifications would require precise monitoring and a deep understanding of Titan's systems to avoid unintended ripple effects across its landscape.

Preserving Titan's methane lakes and hydrological cycle is also key to respecting its scientific value. Titan's methane cycle—a parallel to Earth's water cycle—offers insights into atmospheric processes that operate under conditions vastly different from those on Earth. Drastically warming Titan would likely disrupt this cycle, leading to the rapid evaporation of methane lakes and a fundamental shift in its weather patterns. By retaining Titan's cold temperatures and methane-based hydrology, we preserve its role as a natural analog to Earth, allowing scientists to study how weather, erosion, and landscape formation might function in methane-rich conditions. This knowledge could be essential for understanding other bodies in the universe that may have similar characteristics.

There's also the ethical dimension of transforming Titan, a moon that may contain its own forms of prebiotic chemistry or even subsurface microbial life. Any large-scale changes we impose risk wiping out potential native life forms, or at the very least, altering the environment in a way that obscures their presence. By prioritizing preservation and focusing on adaptable, localized modifications, humanity can explore Titan's potential for habitability while respecting the integrity of its natural systems.

In pursuing a balanced approach to Titan, we confront a larger question: how much should we change an alien world to suit our needs, and at what cost? Rather than trying to mold Titan into a replica of Earth, humanity's challenge may be to adapt itself to this frozen, mysterious moon, learning to coexist with its strange landscape rather than overpower it. By carefully considering the preservation of Titan's natural features and emphasizing adaptability over transformation, we can explore Titan's potential responsibly, leaving its distinctive qualities intact for future generations and for the continued advancement of planetary science.

Chapter 6: The Potential for Life on Titan

The question of life beyond Earth often leads scientists to search for environments that resemble our own, yet Titan challenges this approach with its extreme, alien conditions. While the frigid temperatures, lack of oxygen, and methane-rich atmosphere would seem inhospitable for life as we know it, researchers have begun to explore whether Titan's environment could support forms of life adapted to these unconventional conditions. Titan's unique chemical makeup and atmospheric processes open the door to scientific theories that imagine life in forms that diverge radically from the biology of Earth.

One of the most intriguing hypotheses comes from researchers at Cornell University, who proposed a model for methane-based life that could theoretically thrive in Titan's liquid methane lakes. On Earth, life relies on water as a solvent—water molecules break down and transport nutrients, supporting a vast range of biochemical reactions.

However, Cornell's researchers suggest that life doesn't necessarily require water; under Titan's cold conditions, methane could act as a solvent in place of water. Liquid methane, though chemically inert compared to water, might support a different kind of life that uses hydrocarbons as the foundation for its cellular structures and biochemical processes.

To explore this idea, Cornell scientists devised a hypothetical methane-based cell, which they named an "azotosome." Unlike Earthly cells with membranes composed of phospholipids (molecules suited for a water-rich environment), these hypothetical azotosomes would consist of nitrogen, carbon, and hydrogen compounds—elements abundant on Titan. In simulations, these azotosomes showed promising stability and flexibility, suggesting that such structures could endure Titan's cold and resist decomposition in liquid methane. This model represents a theoretical blueprint for how cellular life might look on a world as alien as Titan, where organic molecules behave differently from those on Earth.

If such methane-based life forms exist, they would represent a paradigm shift in our understanding of biology. They would function without oxygen, drawing energy from chemical reactions involving methane, acetylene, and other hydrocarbons abundant in Titan's atmosphere and lakes. This hypothetical Titanian biology would rely on a metabolic process that's likely very slow, constrained by the low temperatures that limit the speed of chemical reactions. While these life forms would not resemble the active, fast-paced organisms of Earth, they could exist in a state of metabolic stasis, gradually accumulating energy and processing nutrients over long timescales. Such a discovery would redefine the limits of life, showing that biology can adapt to conditions previously thought to be barren.

The implications of this methane-based life model extend beyond Titan. If life can evolve to thrive in methane rather than water, it opens the possibility of finding life on other celestial bodies with similar hydrocarbon environments. Planets or moons once dismissed as too cold or too alien could, in fact, harbor unique ecosystems that follow an entirely different biochemistry. Discovering life on Titan, or even simply confirming the plausibility of methane-based life, would revolutionize our approach to astrobiology, leading scientists to redefine the criteria used to search for habitable worlds.

The idea of methane-based life on Titan also adds a new layer of mystery to the moon's already enigmatic environment. If such life exists, it could be hiding in plain sight within the lakes and seas that cover Titan's polar regions. However, detecting it would require instruments capable of analyzing chemical signatures in Titan's unique atmospheric and liquid conditions—work that NASA's upcoming Dragonfly mission hopes to begin. Even if life remains undiscovered, exploring the biochemical limits within Titan's lakes and atmosphere could still yield groundbreaking insights about the possibilities of life under conditions previously thought to be lifeless.

Cornell's methane-based life theory on Titan invites us to consider that life might not be bound by Earth's constraints. Instead, life could emerge and evolve within environments shaped by radically different chemistry and extreme conditions. Titan's potential as a host for methane-based life reflects the diversity of ways that biology might arise in the universe, urging us to expand our scientific perspective beyond familiar, water-centric models and to imagine the almost unthinkable: life that thrives in an alien sea of methane, under an eternal twilight sky.

Beneath Titan's icy crust lies a hidden ocean, a world shrouded in darkness and mystery, where conditions may be surprisingly conducive to life. Much like Earth's deep oceans, which harbor ecosystems thriving in complete darkness and high pressures, Titan's subsurface ocean could support life forms adapted to an environment cut off from sunlight. This ocean, believed to lie beneath roughly 100 kilometers of water-ice crust, is likely a mix of liquid water and ammonia. Ammonia acts as an antifreeze, keeping the water liquid despite the extreme cold—a vital factor that may enable biological processes otherwise impossible on the surface.

Scientists speculate that the conditions in this subterranean ocean could mirror those found around Earth's deep-sea hydrothermal vents, where organisms draw energy from chemicals rather than sunlight. On Earth, these deep-ocean ecosystems flourish with life, from bioluminescent fish to bizarre bacteria, all reliant on chemosynthesis—extracting energy from chemical reactions involving elements like sulfur and methane. Titan's ocean might host similar life forms if it contains hydrothermal activity or other chemical sources of energy that could sustain microorganisms in a lightless world. Such ecosystems could rely on Titan's natural reserves of methane and other hydrocarbons, or perhaps even trace minerals dissolved from the moon's rocky core.

The existence of life in Titan's hidden ocean is a tantalizing possibility, but confirming it requires exploration tools capable of penetrating its thick ice shell and analyzing the ocean's chemical makeup. This is where NASA's Dragonfly mission enters the picture. Scheduled for launch in the mid-2030s, Dragonfly is a rotorcraft lander designed to explore Titan's surface and gather data on its atmosphere, surface composition, and organic chemistry. While Dragonfly won't drill into the crust to reach the ocean directly, its investigations could still reveal telltale signs of subsurface activity and offer clues to the potential for life below.

Dragonfly's instruments will enable it to analyze Titan's surface and atmospheric composition with unprecedented detail. By examining the chemical reactions occurring on the surface and within the atmosphere, scientists hope to gain insights into how organic molecules behave and accumulate over time in Titan's unique environment. If these surface interactions show complex organic compounds resembling prebiotic chemistry—similar to the chemical processes that once gave rise to life on Earth—it could indicate that Titan's ocean might host even more intricate biological structures. Additionally, Dragonfly's search for chemical traces in surface materials could reveal whether any organic compounds from the subsurface ocean have seeped up through cracks in the icy crust, providing indirect evidence of activity below.

One of Dragonfly's key goals is to search for biosignatures, chemical signs that could suggest the presence of life, past or present. By examining areas where the crust may be thin or where cryovolcanic activity has brought subsurface materials to the surface, Dragonfly could potentially uncover molecules that only form in the presence of biological activity. While discovering life outright may be unlikely with Dragonfly's current mission scope, any signs of complex organic chemistry would strongly support the theory that life could exist within Titan's hidden ocean.

If Dragonfly finds hints of such chemistry, it would fundamentally alter our approach to exploring Titan. A future mission might be tasked with drilling through the icy shell or deploying a submersible probe to dive into the ocean directly, a monumental step toward uncovering life in Titan's depths. The confirmation of microbial life or even prebiotic activity on Titan would redefine our understanding of biology and its adaptability, showing that life can arise in environments far removed from Earth's warm, sunlit oceans.

The subsurface ocean on Titan represents a world of possibilities, a place where life might exist in its most alien forms, thriving in complete darkness, supported by chemical interactions we're only beginning to understand. Dragonfly's mission marks the first significant step toward exploring this hidden ocean indirectly, opening the door to discoveries that could reshape our understanding of life's potential in the universe and point to Titan as one of the most intriguing destinations for future exploration. Through its findings, humanity might come closer to answering one of the most profound questions: Can life, in forms unknown to us, exist in places we've only begun to imagine?

Chapter 7: NASA's Dragonfly Mission – A New Era of Exploration

NASA's Dragonfly mission is an ambitious exploration endeavor aimed at unlocking the secrets of Titan, Saturn's largest moon. Set to launch in the mid-2030s, Dragonfly will be a rotorcraft lander—a unique, drone-like vehicle capable of flying over Titan's surface to explore multiple sites of scientific interest. Unlike traditional rovers limited to ground travel, Dragonfly's design as a multi-rotor drone allows it to navigate Titan's diverse and rugged terrain, from sandy dunes to icy plains, and even potentially the shores of methane lakes. This mobility gives Dragonfly an unparalleled ability to access varied environments, gathering insights across a range of Titan's geological and atmospheric features.

The mission's primary objective is to investigate Titan's prebiotic chemistry and potential habitability. Dragonfly will collect and analyze samples of surface materials, focusing on regions where organic compounds are abundant. Titan's atmosphere and surface contain complex organic molecules, formed from nitrogen and methane reactions under the influence of solar radiation and cosmic rays. By sampling these compounds, Dragonfly aims to study how organic chemistry might naturally evolve in an environment vastly different from Earth's, providing clues to the building blocks that could lead to life.

In addition to sampling Titan's organic-rich surface, Dragonfly will analyze atmospheric and surface interactions. Using a suite of scientific instruments, it will examine the composition of Titan's atmosphere and surface materials, measuring the chemical properties and possible energy sources that could sustain biological activity. Dragonfly's instruments are designed to detect biosignatures or patterns in organic molecules that could indicate biological processes. This data could be crucial for understanding whether Titan's chemistry resembles the early conditions that once set life in motion on Earth, making Titan a potential analog for studying prebiotic environments.

Dragonfly's drone capabilities are central to its mission success. With its rotors, Dragonfly can cover substantial distances—up to tens of kilometers in a single flight—far beyond the range of traditional landers and rovers. This agility enables Dragonfly to sample and study different types of terrain and surface features, from dunes and plains to impact craters and possibly even regions where cryovolcanic activity has brought subsurface materials to the surface. Titan's low gravity and dense atmosphere, which is four times thicker than Earth's, make flight an efficient method of travel, allowing Dragonfly to glide easily over Titan's surface, conserving energy and maximizing the mission's scope.

Dragonfly's technology also includes autonomous navigation systems that allow it to adapt to Titan's unpredictable surface. Titan's dense, hazy atmosphere means visibility is limited, so Dragonfly will rely on onboard sensors and mapping systems to assess and adjust its path as it flies, ensuring safe landings on varied and potentially uneven terrain. This autonomous capability allows Dragonfly to explore complex landscapes that would otherwise be out of reach, enhancing its ability to gather samples from some of Titan's most scientifically promising areas.

The mission also aims to explore Titan's seasonal and atmospheric changes. By observing how Titan's atmosphere and surface interact over time, Dragonfly will provide valuable data on the moon's methane cycle—an analog to Earth's water cycle—and on how this cycle shapes Titan's landscape and climate. Insights from these observations will deepen our understanding of Titan as an evolving world, with active geological and chemical processes that may resemble early Earth.

Dragonfly's mission represents a pioneering approach to planetary exploration, using drone technology to access and study regions that traditional rovers cannot reach. Through its sample collection and chemical analysis, Dragonfly aims to unlock the secrets of Titan's organic-rich environment and potential for life, advancing our understanding of what makes a world habitable. If successful, Dragonfly could reveal a world where organic chemistry has taken a unique path, shedding light on how life might arise and adapt in the most unexpected corners of our universe.

Dragonfly's mission to Titan opens up the possibility of utilizing Titan's unique resources to extend or even support long-term exploration. One of the most intriguing proposals is to use Titan's abundant methane lakes as a fuel source. Titan's lakes and seas contain liquid methane and ethane, hydrocarbons that could be harvested and converted into fuel for future missions. This refueling capability would mean that instead of relying entirely on resources brought from Earth, an exploration vehicle or even a base on Titan could use the moon's natural reserves to stay operational, a concept known as in-situ resource utilization (ISRU). Extracting and processing methane for energy would represent a critical step in establishing self-sufficiency in space exploration, reducing the logistical challenges and costs of constant resupply from Earth.

Alongside methane as a fuel source, researchers have also explored the possibility of extracting oxygen from Titan's surface ice. Titan's crust is composed mainly of water ice, and by splitting water molecules into hydrogen and oxygen, oxygen could be used as a component in a breathable atmosphere or combined with methane as a powerful oxidizer in fuel combustion. Such a system would require advanced technology capable of operating in Titan's frigid conditions, but the potential benefits are significant. Accessing oxygen from the ice would allow for sustainable fuel production, especially valuable if a human presence on Titan were ever established, as it would enable breathable air and rocket fuel for missions traveling beyond Titan.

In terms of mission planning, Dragonfly is set to launch in 2027, with arrival on Titan expected around 2034 due to the vast distance between Earth and Saturn. Once there, Dragonfly will begin its exploration, covering multiple locations over several Titanian years (each year on Titan lasts about 29 Earth years). During this time, it will gather samples, analyze the atmospheric composition, and provide insights into Titan's organic chemistry and surface interactions. Given the extensive time required for data transmission and the need for careful study across different sites, the Dragonfly mission is designed to span several years, allowing scientists to build a comprehensive picture of Titan's surface and atmospheric dynamics.

Beyond Dragonfly, the future of Titan exploration holds ambitious possibilities. One proposed follow-up mission involves sending a submarine to explore Kraken Mare, Titan's largest sea, which is filled with liquid methane and ethane. This submarine, if realized, would represent an unprecedented step in interplanetary exploration, designed to navigate the frigid depths of an alien sea to study its chemical composition, possible currents, and even the potential for unique forms of life. A Titan submarine would need to operate autonomously, navigating underwater with sensors equipped to analyze Titan's methane-rich liquid and capable of withstanding pressures at depths over 300 meters. This type of mission would push technological boundaries, requiring new adaptations in underwater robotics and chemistry analysis in hydrocarbon environments.

If implemented, the Titan submarine mission could launch by the 2040s, depending on advancements in exploration technology and the outcomes of Dragonfly's findings. Should Dragonfly uncover hints of complex organic chemistry or other signs that Titan's seas and lakes hold valuable data on prebiotic or even biotic processes, interest in a submarine mission would likely accelerate. This would mark a milestone in humanity's exploration of oceanic environments beyond Earth, offering an unprecedented glimpse into the depths of an alien sea, where conditions challenge our understanding of planetary science and astrobiology.

Together, these missions—the drone-enabled Dragonfly and the proposed Titan submarine—represent the cutting edge of planetary exploration, utilizing innovative resource management and navigation technologies that could redefine what's possible in space exploration. By exploring Titan's diverse landscapes and unique chemical resources, these missions aim to uncover not only the geological and chemical secrets of this icy moon but also the practical pathways for sustaining future exploration in the outer solar system. Through these advances, humanity moves closer to the vision of exploring and even living in environments beyond our world, empowered by Titan's own resources.

Chapter 8: Titan as Humanity's Next Home – Infrastructure and Resource Utilization

Titan's abundant hydrocarbons—methane and ethane—present a unique resource for prospective human settlers, not only as potential fuel but also as foundational building materials. In Titan's frigid climate, hydrocarbons remain in a stable liquid or solid state, forming lakes, rivers, and dunes that could serve as reservoirs for essential resources. These hydrocarbons can be refined into different forms: methane could power fuel cells or engines, while solid hydrocarbons might be processed into plastics and other structural materials. This versatility in resource use could enable a self-sustaining approach to constructing habitats, tools, and infrastructure on Titan, reducing the need to transport supplies from Earth.

For energy production, hydrocarbons provide an essential advantage in Titan's low-sunlight environment. Methane, in particular, could fuel energy systems similar to those on Earth, powering heating systems and generating electricity for habitats and equipment. Titan's hydrocarbon reserves are virtually limitless on a human scale, providing energy security in an environment where traditional sources are constrained. This locally sourced energy could support long-term exploration, enabling a stable presence on Titan and opening the door for larger-scale missions to Saturn's other moons and beyond.

Titan's liquid methane rivers also offer an opportunity for alternative, sustainable energy production. Hydroelectric power, driven by the slow flow of methane rivers and streams, could be harnessed to generate electricity. By positioning small-scale turbines within these methane channels, explorers could produce a steady power supply without consuming hydrocarbons directly. While Titan's rivers may not flow with the speed or volume of water rivers on Earth, their continuous movement and chemical stability provide a potential for continuous, if modest, energy production. This source could be especially valuable in remote areas or as a supplementary energy source, reducing dependence on fuel combustion and ensuring that power remains available even if methane-based fuel resources are limited or need to be conserved.

However, Titan's dim sunlight poses a significant limitation for solar power. At approximately 1.5 billion kilometers from the Sun, Titan receives only about one percent of the sunlight Earth does. This weak light, filtered further by Titan's thick, hazy atmosphere, means that solar panels would struggle to produce substantial power. Solar energy could potentially play a role in Titan's equatorial regions or as a minor supplement for low-power applications, but it would not be sufficient as a primary energy source for any extensive human activity. The limited solar power available highlights the necessity of developing other methods, such as hydrocarbon-based systems and hydroelectric turbines, to support life and exploration on Titan.

By leveraging Titan's rich hydrocarbon landscape and developing infrastructure for sustainable energy, future missions and habitats could establish a more independent foothold on this icy moon. Combining methane-based fuels, hydroelectric power from Titan's rivers, and possibly even geothermal energy from cryovolcanic sites would create a multifaceted energy strategy, empowering explorers to navigate and thrive in Titan's extreme environment without relying heavily on Earth-supplied resources. This approach embodies the vision of self-sufficiency in space, using the bounty of an alien world to fuel human ingenuity and survival in one of the most remote places in our solar system.

Establishing agriculture on Titan, while challenging, is a possibility that could greatly enhance the sustainability of a human presence on this distant moon. Titan's surface, rich in nitrogen, presents a unique resource for cultivating soil that could support plant life. Nitrogen is a key element for plant growth, integral to proteins and chlorophyll, and Titan's nitrogen-rich atmosphere provides an abundant supply. However, due to the absence of oxygen and water, and the extreme cold, this nitrogen would require careful processing, likely mixed with imported organic material and minerals to create a soil that could support plant roots and nutrient absorption.

To facilitate growth in Titan's low-light environment, agriculture would depend on artificial sunlight within greenhouse enclosures. These greenhouses, pressurized to Earth-like conditions and warmed to suitable temperatures, would replicate an Earth-like growing environment while shielding plants from Titan's cold and unbreathable air. The greenhouses could use LED lighting optimized for photosynthesis, allowing for year-round growth regardless of Titan's dim sunlight and long seasonal cycles. Given the efficiency of modern LED technology, this setup could provide a reliable energy-to-yield ratio, powered by Titan's methane resources or hydroelectric systems, creating a closed-loop system for sustaining plant growth.

Greenhouses on Titan would also benefit from a controlled environment where carbon dioxide and oxygen levels could be tailored to maximize crop yield. Titan's atmospheric nitrogen could be directly harnessed to fertilize the soil within these greenhouses, reducing the need for synthetic fertilizers and making the process more sustainable. By cultivating crops within insulated, artificially lit structures, future explorers could produce fresh food, reduce dependency on Earth-supplied provisions, and recycle the oxygen generated by plants to enrich breathable air within habitats. This approach could support a wide range of crops, from staple grains to nutrient-rich vegetables, fostering a level of self-sufficiency that would be invaluable for long-term missions.

Beyond agriculture, conceptual designs for habitats on Titan would also lean heavily on the moon's natural resources. One possible habitat design involves using Titan's hydrocarbon sands to construct pressurized domes or shelters. These hydrocarbons, processed into durable plastics, could provide the building materials for lightweight, insulated structures. By incorporating layers of plastic insulation, habitats could retain warmth and provide robust protection against Titan's extreme cold. Such structures could be modular, allowing explorers to expand the habitat as needed while reinforcing it with locally sourced materials, thereby reducing the dependence on materials imported from Earth.

In addition to plastic-based structures, habitats on Titan could incorporate ice from Titan's crust, which, when properly insulated, would serve as a natural building material. Thick walls of ice could be reinforced with structural elements derived from hydrocarbon polymers, providing a stable, insulated barrier between the interior of the habitat and Titan's frigid environment. These ice-and-polymer composites would create a thermal barrier, keeping the inside of habitats warm while using Titan's natural landscape as the core building material.

These habitats would likely feature interconnected modules—living quarters, laboratories, greenhouses, and storage areas—connected by insulated walkways that would allow safe travel between sections. Each module would be equipped with systems to monitor and regulate temperature, air quality, and radiation levels, creating a controlled micro-environment within Titan's extreme landscape. Additionally, methane-powered heating and energy systems could sustain warmth and light, while advanced oxygen-recycling systems would ensure a steady supply of breathable air.

By utilizing greenhouses for food production and constructing habitats with Titan's abundant hydrocarbons and water ice, future settlers could achieve a level of self-reliance that minimizes the need for continuous Earth-based resupply. This approach to habitat design and agriculture embodies the concept of in-situ resource utilization, transforming Titan's alien resources into the essentials for life. In leveraging Titan's native elements, humanity could forge a livable environment, pioneering a way of life adapted to the unique conditions of Saturn's distant moon, creating a foothold in one of the most remote and resource-rich environments within our solar system.

Chapter 9: The Ethical and Practical Implications of Colonizing Titan

The idea of colonizing Titan—one of the few places in our solar system that might harbor conditions suitable for life, even if vastly different from life on Earth—presents significant ethical questions. As humanity pushes the boundaries of exploration and dreams of extending its presence beyond Earth, it must also confront the potential consequences of altering an alien world. Titan's distinctive chemistry and possible subsurface ocean make it a scientifically precious environment, and our arrival could irreversibly change this ecosystem. This raises a fundamental question: should we prioritize exploration and colonization, or should we respect and preserve Titan as a natural entity, potentially untouched by Earth-based life?

A key concern is the possibility that Titan already hosts microbial or prebiotic life, perhaps adapted to its methane lakes or deep, ammonia-rich ocean. If this is the case, introducing Earth-based organisms could have serious implications. Terrestrial microbes, no matter how small or seemingly harmless, might disrupt Titan's native chemical processes or even outcompete potential local life forms for resources. Such biological contamination could alter Titan's natural cycles, erasing evidence of any unique, methane-based life forms and replacing it with a simplified ecosystem dominated by Earth-origin microbes. This risk highlights the need for extreme caution; any introduction of Earth organisms, even unintentionally, could mean the loss of an invaluable scientific and philosophical opportunity to understand life as it could exist independently from Earth's ecosystem.

Another ethical dimension lies in the environmental impact of Earth-based activities on Titan's delicate atmosphere and surface. Titan's landscape—its methane lakes, hydrocarbon dunes, and cryovolcanoes—represents an alien environment shaped over eons by processes unknown on Earth. Mining Titan's methane, constructing habitats, and establishing agricultural systems could disrupt this delicate balance, potentially triggering chemical changes that would alter Titan's landscape, atmosphere, and methane cycle. Future settlers might find it difficult to avoid altering Titan's surface through energy production, greenhouse installations, or resource extraction, creating an inevitable footprint that could transform Titan's unique conditions permanently.

Introducing Earth-based organisms to Titan's ecosystem could also pose risks to human explorers. Microbes, plants, and even larger organisms adapted to Earth's oxygen-rich environment might fail to survive in Titan's frigid, oxygen-poor atmosphere, requiring special measures to contain and protect them. However, in the event that any organisms adapt unexpectedly to Titan's conditions, they could mutate in ways that make their behavior or impact unpredictable. This "biological drift" could result in organisms evolving to utilize Titan's native resources or even interfering with critical human infrastructure, such as methane-based energy systems, in unforeseen ways.

The introduction of agriculture, with the use of greenhouses and artificially heated environments, could further exacerbate the risk of contamination, not only introducing non-native species but potentially creating microclimates where invasive organisms might flourish. While these microenvironments would be contained, any breaches or failures could spill Earth's organic material onto Titan's surface, adding terrestrial matter to an environment that might not have evolved to handle it. This raises a question about stewardship: how much do we, as human explorers, have the right to reshape Titan, and what measures should we take to minimize our environmental impact?

In light of these ethical and environmental concerns, some scientists and ethicists advocate for a "preservation first" approach to Titan exploration. This would mean developing and enforcing strict planetary protection protocols, ensuring that all equipment and habitats are rigorously sterilized before deployment to minimize contamination risks. More radical approaches suggest prioritizing robotic missions to gather as much data as possible without human presence, reserving direct human exploration for when we have a more comprehensive understanding of Titan's environment and any potential biosphere.

Ultimately, the ethical considerations surrounding Titan colonization challenge humanity to think deeply about our responsibilities as explorers. Preserving the moon's natural state—particularly if it may harbor life—requires that we weigh our scientific curiosity and colonization ambitions against the intrinsic value of Titan's untouched ecosystem. While Titan offers a promising frontier for exploration and human settlement, it also serves as a reminder of the complex responsibilities that come with reaching beyond Earth. Balancing our aspirations to explore and thrive with our duty to preserve and respect alien worlds may define the future of interplanetary exploration, compelling us to consider the legacy we leave as we venture into new and uncharted realms.

Long-term colonization of Titan presents formidable technical and financial challenges that demand innovation, investment, and careful planning. Unlike Mars, which is relatively close and better understood, Titan sits over a billion kilometers from Earth, requiring new engineering approaches and solutions to sustain a human presence in its extreme, methane-rich environment. The physical and logistical obstacles alone are immense: supporting human life in Titan's subzero climate, creating sustainable food and energy systems, and protecting habitats from frigid temperatures and potential contamination—all while maintaining safe travel and communication over vast distances.

One of the primary technical obstacles is Titan's distance from Earth, which complicates transportation and resupply. A round-trip mission to Titan could take close to 15 years, including travel and mission execution time. This delay means that any equipment, supplies, or personnel would need to function independently for extended periods, demanding an unprecedented level of reliability in life-support systems, habitats, and energy infrastructure. To solve this, future missions will likely rely on in-situ resource utilization (ISRU), tapping into Titan's local resources—such as hydrocarbons for fuel and water ice for oxygen—to minimize the need for Earth-based supplies. However, developing technology that can convert these resources into usable forms will require costly research and rigorous testing.

Power generation on Titan poses another critical challenge. With limited sunlight reaching Titan's surface, solar power is impractical, pushing the focus onto methane-based energy systems or even nuclear reactors. While these alternatives are promising, each brings its own set of technical hurdles. Methane-fueled systems would need to be highly efficient in converting Titan's abundant hydrocarbons into energy, while nuclear reactors would require extensive shielding, security protocols, and waste disposal strategies. Both systems would also require careful maintenance and redundancy to ensure a continuous power supply in Titan's remote and harsh conditions.

The financial demands of a Titan colony are equally significant. Building infrastructure that can withstand extreme cold, developing advanced life-support and energy systems, and planning for long-duration missions all come at an enormous cost. Current estimates for similar, albeit smaller-scale missions, such as those proposed for Mars, already range in the tens of billions of dollars, suggesting that a Titan mission could easily exceed these figures. Funding such a venture would likely require contributions from multiple nations, private companies, and perhaps even international space agencies pooling resources. Securing long-term financial support also means justifying the mission's scientific and exploratory value to governments and stakeholders, who will need to be convinced of Titan's importance as a site for research, resource exploration, or future habitation.

Given these technical and financial obstacles, international cooperation will be essential. A mission to Titan could serve as a unifying project for spacefaring nations, pooling expertise, technology, and funding to make such an ambitious endeavor feasible. Collaborative efforts, like those seen in the International Space Station, illustrate the potential for shared missions that draw on diverse resources and expertise from around the world. This approach would not only distribute costs but also promote the exchange of ideas and technologies, enhancing innovation and reducing redundancy across the project's many components.

Long-term planning, including creating a phased approach to exploration and development, will be crucial for success. Initial robotic missions like

Dragonfly could establish a foundation of knowledge, laying the groundwork for future, more extensive missions. Phases might include establishing robotic infrastructure for resource processing and power generation, followed by small-scale human missions to test life-support systems and habitat viability. Such gradual, structured steps would allow for in-depth testing of essential systems, creating a safer environment for larger, sustained human operations down the line.

Furthermore, shared governance and regulatory frameworks would be necessary to manage Titan's exploration responsibly. This includes ethical considerations, such as preventing biological contamination and respecting Titan's potential for indigenous microbial life. International agreements, similar to the Outer Space Treaty, could guide the protection of Titan's environment and ensure that colonization proceeds with transparency, accountability, and respect for scientific integrity.

In facing the monumental task of Titan colonization, humanity must combine technological innovation, cooperative spirit, and a shared vision for the future. Such an endeavor challenges us not only to solve practical problems but to create a model of exploration that values sustainability, mutual benefit, and ethical stewardship. By working together, pooling resources, and committing to a long-term plan, we can take meaningful steps toward realizing the dream of reaching and inhabiting distant worlds, leaving a legacy of cooperation and discovery that serves as a foundation for future generations.

Chapter 10: The Future of Human Exploration – Titan and Beyond

Titan represents a bold frontier in humanity's pursuit of space exploration and interplanetary colonization. While Mars and Earth's Moon remain the primary candidates due to their proximity, Titan offers unique advantages that make it a promising, if more challenging, long-term target. In many ways, Titan complements and even surpasses these more immediate options, promising a blend of habitability features that position it as a valuable next step for humanity's deeper ventures into the solar system.

Unlike Mars and the Moon, which have thin or negligible atmospheres, Titan is wrapped in a dense, nitrogen-rich atmosphere—1.5 times the atmospheric pressure at Earth's surface. This pressure would enable humans to walk on Titan's surface without pressurized suits, needing only insulated thermal protection and oxygen supplies. The atmosphere also serves as a natural barrier against cosmic and solar radiation, a protection that neither Mars nor the Moon provides. Mars, with its thin atmosphere, requires significant radiation shielding, and the Moon's lack of atmosphere exposes it entirely to the harshness of space. Titan's thick atmosphere and Saturn's powerful magnetosphere create a uniquely protective environment, ideal for establishing long-term colonies that prioritize safety from radiation.

Titan's abundant natural resources also set it apart. Its surface hosts vast lakes and rivers of liquid methane and ethane, which, despite being toxic to Earth-based life, are valuable for generating energy and producing fuel. The methane could be harvested and used in energy systems to heat habitats, power machinery, or fuel return journeys. Mars and the Moon, in contrast, lack such accessible hydrocarbon reserves. While Mars has frozen water and mineral resources and the Moon possesses some water ice, neither provides the abundant, ready-to-use fuel that Titan's hydrocarbons offer. This opens up possibilities for establishing energy independence on Titan, harnessing local resources to build a self-sufficient base far removed from reliance on Earth.

The low surface gravity and dense atmosphere on Titan also allow for unique transportation options, such as drone-like crafts or gliders, to explore the moon's diverse terrains, which include dunes, mountains, and methane lakes. This ease of aerial exploration would be invaluable for scientists and explorers aiming to study different regions of the moon quickly and safely. Mars, with its thin atmosphere, cannot easily support such aerial vehicles, and the Moon's lack of atmosphere means flight is not an option. Titan's atmosphere thus allows for innovative, energy-efficient exploration techniques that further enhance its suitability for sustained scientific study.

One major challenge is Titan's distance from Earth. Over a billion kilometers away, it requires a multi-year journey, making resupply missions costly and infrequent. This means any mission to Titan would need to be almost entirely self-sustaining, with robust life-support and resource extraction systems in place before a long-term human presence could be considered. Titan's remoteness thus positions it as a goal for the future, after humanity has gained experience and technological advancements from closer colonies on Mars or the Moon. With successful bases established there, we could use these colonies as stepping stones to reach Titan, possibly even sending advanced robotic missions to prepare the moon for human arrival.

While Mars and the Moon offer immediate testing grounds for off-world settlement, Titan represents a more ambitious target, requiring the adaptability and resourcefulness needed to thrive in an environment profoundly different from Earth. Its thick atmosphere, natural radiation shielding, and abundant hydrocarbons could enable the development of a colony that supports human life with minimal reliance on Earth. A Titan colony would embody the ethos of adaptation and sustainable resource utilization, serving as a powerful example of how humanity can integrate with an alien environment rather than reshape it entirely.

In the context of humanity's broader ambitions, Titan aligns with our evolving vision of becoming a spacefaring species. Unlike Mars or the Moon, where colonization may involve creating Earth-like conditions, Titan encourages us to think beyond familiar ecosystems, fostering the skills and technologies to thrive in profoundly different environments. As humanity looks toward a future among the stars, Titan offers a glimpse of what it might take to truly live beyond Earth—learning to harness alien resources, embrace environmental differences, and build communities capable of surviving on their own. This unique potential situates Titan as a key milestone in our journey into the cosmos, a frontier that challenges us to redefine the very concept of habitability in our quest to explore and understand the universe.

The prospect of colonizing Titan, Saturn's largest and most complex moon, pushes the boundaries of human ambition and technological ingenuity. While Titan's thick atmosphere, natural resources, and protective magnetosphere make it an attractive candidate for exploration and potential colonization, reaching and establishing a self-sustaining human presence on this distant moon will demand a series of groundbreaking advancements in technology and mission design. Future missions and research are poised to lay the groundwork, gradually transforming the dream of Titan colonization from science fiction into an achievable reality.

Initially, robotic missions like NASA's upcoming Dragonfly will lead the way. Dragonfly, a rotorcraft lander capable of exploring Titan's surface and atmosphere, will serve as a pioneering mission, collecting essential data on Titan's organic compounds, surface composition, and environmental conditions. This mission is designed to operate autonomously and cover diverse terrains, providing insight into the practicality of future human operations on Titan. Data gathered by Dragonfly will inform us about the behavior of methane lakes, the stability of Titan's icy crust, and the potential for using Titan's hydrocarbon resources for energy. These initial robotic missions are invaluable, laying the foundation for larger, more complex missions by identifying both challenges and possibilities.

As interest in Titan grows, future missions will likely involve deploying advanced robotic infrastructure that can extract and process resources autonomously, a vital step for any long-term settlement. In-situ resource utilization (ISRU) technology will be key—systems capable of harvesting methane from Titan's lakes, extracting oxygen from water ice, and generating energy from these resources will be essential to sustaining life on

Titan. Refining methane and combining it with oxygen could create fuel for return journeys or further exploration into the outer solar system, making Titan a possible outpost for even deeper space missions. These technologies would not only support human survival but also allow explorers to live and work in an environment where traditional resupply is logistically unfeasible.

Future advancements will also include habitat designs that can withstand Titan's extreme cold, atmospheric pressure, and volatile chemistry. Habitats will need to be well-insulated, heated, and possibly built from materials derived from Titan's own hydrocarbons, which could be processed into durable polymers. By incorporating materials sourced from Titan itself, these habitats could become self-sustaining structures that require minimal Earth-based resources. Technologies that recycle air, water, and energy will need to reach new levels of efficiency, creating closed-loop life-support systems that can function independently for extended periods. Artificial lighting systems, optimized for plant photosynthesis, would be necessary to sustain agriculture in greenhouses, allowing settlers to produce their own food and achieve a degree of self-sufficiency in food production.

The success of Titan colonization also depends on advancements in transportation and propulsion. Given the enormous distance from Earth to Titan, spacecraft propulsion systems will need to become faster, more efficient, and capable of carrying heavier payloads. Nuclear propulsion, solar sails, or advanced ion thrusters are among the technologies that could reduce travel time, enable reliable resupply, and facilitate the transport of larger equipment and personnel. Improved communication infrastructure, possibly utilizing satellites positioned between Earth and Saturn, will also be essential to maintaining reliable contact and coordination with Titan colonies, despite the lengthy transmission delays inherent in deep-space communication.

As these technological hurdles are addressed, humanity's quest to reach Titan reflects a broader philosophical journey—a desire to extend our presence beyond Earth and into the outer reaches of our solar system. Titan stands as a symbol of human resilience and adaptability, urging us to imagine a future where we can thrive in conditions far removed from our home planet. Unlike the Moon or Mars, where colonization strategies often focus on replicating Earth-like environments, Titan challenges us to innovate in ways that harmonize with its unique atmosphere and resources. It represents a chance to develop technologies and cultural philosophies that embrace, rather than resist, the distinctiveness of an alien world.

In this broader context, Titan symbolizes a new frontier in our understanding of what it means to be a spacefaring species. It invites us to test the limits of resourcefulness, to create closed-loop systems, to forge self-reliant colonies, and to cultivate a mindset that sees value in the unfamiliar. By rising to these challenges, humanity can transform Titan from a distant curiosity into a model for future planetary settlements across the solar system and beyond. This journey to Titan represents a defining chapter in humanity's quest to transcend Earth's boundaries, fostering a spirit of exploration that may one day take us to the stars.

Conclusion

Titan stands as a beacon of both mystery and possibility, a place where the familiar and the alien converge in a way that captivates and challenges humanity. This distant moon, with its thick atmosphere, liquid methane lakes, and hidden ocean, presents a world that is neither Earth-like nor entirely inhospitable. It is a realm that stretches the imagination, demanding that we rethink what life, habitability, and exploration could mean in conditions so different from our own. Titan's unique environment, rich in natural resources and potential, places it at the frontier of scientific discovery and the edge of our understanding of where and how life might exist.

For scientists, Titan is a treasure trove of mysteries, from the possibility of methane-based life in its lakes to the potential for unknown organisms thriving in its subterranean ocean. Its chemistry may hold secrets about prebiotic processes, offering clues to how life could arise on worlds vastly different from Earth. Each lake, river, and dune is a piece of a grand puzzle, waiting to reveal how organic compounds behave in an environment shaped by hydrocarbons rather than water. Titan calls us to explore these questions, drawing us closer to understanding not just the nature of life, but the adaptability of life's building blocks.

Titan is also a test of human ambition, resilience, and innovation. It presents challenges that will require every advancement in technology and every ounce of human ingenuity. The extreme cold, the scarcity of breathable oxygen, and the vast distance from Earth demand a new level of self-sufficiency and resourcefulness. Yet these obstacles are also what make Titan a symbol of human capability—the chance to show that we can adapt to an environment so foreign that it pushes the boundaries of our engineering, science, and ethics. In its remoteness and complexity, Titan represents not just a goal for the future, but a journey in itself, a journey that compels us to go further than we have ever gone before.

The allure of Titan is rooted in both its promise and its peril. It holds the potential to serve as a new home or outpost, a self-sustaining community that can live off its resources and foster new kinds of human endeavor. Yet, Titan also reminds us of the responsibility we carry as we venture beyond Earth. The possibility of contaminating or disrupting an ecosystem that may already harbor life calls for caution and respect. This balance—between discovery and preservation, ambition and restraint—will define our relationship with Titan and worlds like it.

As humanity steps out further into the solar system, Titan embodies the spirit of exploration, a reminder of why we reach for the stars. It challenges us to learn, adapt, and grow, to look beyond the familiar and embrace the unknown. Titan is not merely a destination; it is a testament to human curiosity and our capacity to meet the extraordinary with courage and imagination. In this journey to Titan and beyond, we stand on the cusp of a new era, one where each step takes us farther from Earth and closer to understanding our place in the cosmos.

www.ingramcontent.com/pod-product-compliance
Lightning Source LLC
Chambersburg PA
CBHW070248220526
45465CB00004B/1561